Maple V Release 3
Notes

Maple V Release 3 Notes

Brooks/Cole Publishing Company

An International Thomson Publishing Company

Pacific Grove   Albany   Bonn   Boston   Cincinnati   Detroit   London   Madrid
Melbourne   Mexico City   New York   Paris   San Francisco   Singapore   Tokyo
Toronto   Washington

Maple V Release 3 Notes © 1995 by Waterloo Maple Software. All rights reserved. No part of this work may be reproduced, stored in a database or retrieval system, or transcribed, in any form or by any means—electronic, mechanical, photocopying, or otherwise—without the prior written permission of the publisher, Brooks/Cole Publishing Company, Pacific Grove, California 93950.

Maple is a registered trademark of Waterloo Maple Software. All other product names mentioned in this work are trademarks or registered trademarks of their respective companies.

Printed in the United States of America

10 9 8 7 6 5

For more information, contact: Brooks/Cole Publishing Company
511 Forest Lodge Road
Pacific Grove, CA 93950 USA

For orders, call: 800-354-9706 Central Time
For technical support:
    call: 800-214-2661 (6:00 A.M.–4:00 P.M. Pacific Time)
    fax: 408-375-0120
    email: SUPPORT@BROOKSCOLE.COM

International Thomson Publishing
Berkshire House 168–173
High Holborn
London WC1V 7AA
England

International Thomson Publishing GmbH
Königwinterer Strasse 418
53227 Bonn
Germany

Thomas Nelson Australia
102 Dodds Street
South Melbourne, 3205
Victoria, Australia

International Thomson Publishing–Asia
221 Henderson Road #05–10
Henderson Building
Singapore 0315

Nelson Canada
1120 Birchmount Road
Scarborough, Ontario
Canada M1K 5G4

International Thomson Publishing–Japan
Kyowa Building, 3F
2-2-1 Hirakawacho
Chiyoda-ku, 102 Tokyo
Japan

Contents

Introduction 1
  Scope of This Guide 1
  How to Use This Guide 1
  Overview of Maple V Release 3 2
    Interface 2
    Mathematics 2
    Graphics 3
    System and Language 3
  Where to Go for More Information 3

Interface 5
  Worksheets 5
  Typographic Output 6
  Help Facilities 7
    Help Topic Browser 7
    New Help Commands 8
    Worksheet Examples 8

Mathematics 9
  Symbolics 9
    The **assume** Facility 9
    Fourier and Laplace Transforms 10
    Integration 11
    Unevaluated Functions 12
    Differential Equations 13
    The **minimize** command 15
    Simplification of Radicals 15
    Algebraic Numbers, Functions, and Fields 18
    Additional Functionality 19
  Numerics 20

Contents

dsolve(,numeric)	20
Additional Functionality	22
Complex Variables	22
Complex Data Type	23
Automatic Complex Arithmetic	23
More Functions Supported	23
Additional Functionality	24
Packages	24
Packages New to Release 2	24
Packages Improved for Release 2	25
Packages New to Release 3	25
Packages Improved for Release 3	27
The Share Library	28

Graphics . . . 31
Contour Plots . . . 31
Implicit Plots . . . 32
Vector Field and Gradient Vector Field Plots . . . 32
Scaling Axes . . . 33
Point Style . . . 34
Line Style . . . 34
Line Thickness . . . 35
Plot Labeling . . . 35
Rendering . . . 35
Lighting Models . . . 35
Coloring . . . 36
Plot Data Structure . . . 37
Other Improvements . . . 37
Animation . . . 37
    Two-Dimensional Animation . . . 38
    Three-Dimensional Animation . . . 38
Output Formats and Devices Supported . . . 38

System and Language . . . 41
Libraries and I/O . . . 41

Contents

- Multiple Library Access ... 41
- Maple Library Archiver ... 41
- Formatted Input and Output ... 42
- **.m** File Format ... 43
- Clearing the Session ... 43
- Error Handling and Debugging ... 43
  - Automatic Type Checking ... 44
  - **errorbreak** ... 44
- Language ... 45
  - Name Protection ... 45
  - Global Variables ... 45
  - String Processing ... 46
- Other Improvements ... 46

Introduction

The third major release of the *Maple V* program, *Maple V Release 3*, provides significant advances in all areas of the computer algebra environment. Not only have the mathematical capabilities of *Maple V* improved, but major enhancements have also been made in *Maple V*'s interface, graphics, system, and language. This guide, *Maple V Release 3 Notes*, highlights many of these additions and improvements.

Scope of This Guide

These notes summarize the changes to *Maple V* since its initial release. Changes specific to *Release 3* are denoted by **R3** in the margin, as shown here.

Although many sections of this guide are written in tutorial style, *Maple V Release 3 Notes* is *not* a beginning level *Maple* tutorial. The *Maple V Flight Manual* and the *Maple V Quick Reference*, both published by Brooks/Cole Publishing Company, provide a thorough introduction to *Maple V*.

We have summarized here those changes that will most affect your use of *Maple V*. Because of the number of additions and improvements made in *Releases 2* and *3*, a complete listing is impracitcal in a guide such as this. For references to more detail, see the section *Where to Go for More Information* on page 3.

How to Use This Guide

Before you start using the *Maple V Release 3* software, quickly read through this guide. If you are new to *Maple*, it will give you an idea of the power of a sophisticated computer algebra system. If you are an established *Maple* user, it will give you an idea of the tools that have been added or improved to make your calculations more efficient. Because segments of *Maple*'s functionality fall naturally into separate categories, the information in this similarly organized into the following chapters:

- Interface
- Mathematics

Overview of Maple V Release 3

- Graphics

- System and Language

If an improvement spans multiple categories, it is listed under the area in which it has the most influence. An index is also provided to help you find the information you need.

Overview of Maple V Release 3

Information on *Release 3* is contained throughout this guide, sometimes in sections of its own, sometimes interspersed with *Release 2* information. The following list provides an overview of the most important improvements in *Release 3*.

Interface

- improvements to the help system, including keyword search and commands to display selected information from the help pages

- a new facility for converting entire *Maple* worksheets into LaTeX documents

- Windows Tool Bars for ease of use

- improved installation program for Windows

- better portability of worksheets across platforms

- subscripts printed as subscripts.

Mathematics

- a completely new and improved statistics package

- the share library, a compendium of user-submitted code, is now much larger and contains, for example, more than 70 sample worksheets

- improved handling of differential equations, including new algorithms for solving linear differential equations and a new unevaluated command **DESol**

- new algorithms for solving numeric differential equations

Introduction

- improved handling of **sqrt** and other radical simplifications
- multivariate polynomial factorization over GF(p^k)
- improved algorithm for Hermite and Smith normal forms and computation of the multiplier matrices
- complex variable evaluation of the Riemann Zeta function
- improvements to indefinite and definite integration
- the value of **signum(0)** definable by the user.

Graphics

- two-dimensional plotting of discontinuous functions
- optional parameters for control of number of contours, line style, line thickness, symbols used for points, and fonts used for text
- the speed of implicit plotting has been significantly increased.

System and Language

- name protection facilities, **protect** and **unprotect**
- new methods of handling of **local** and **global** variables within procedures
- **searchtext** and **SearchText** for searching within strings
- conversion tools, **m2src**, which converts *Release 2* **.m** files into *Release 2* source files, and **updtsrc**, which converts *Release 2* source files to *Release 3* source files
- inclusion of the library archiver, **march**, and the syntax checker, **mint**, with all platforms.

Where to Go for More Information

This guide is a *brief* introduction to the changes made to *Maple V* since its initial release. In many cases, only alterations to existing commands are listed—explanations of unchanged aspects of the commands are omitted. Examples are

Where to Go for More Information

limited to outlining the basic workings of new commands. For more information and detailed examples, refer to the *Maple* help pages, accessible either through the **?<topic>** syntax or through the hierarchal topic browser. See the section *Help Facilities* on page 7 for more details.

Two special help pages, **?updates,v5.2** and **?updates,v5.3**, provide additional information about new facilities in these two releases. If you want more details of functionality and algorithms, read these help pages. They are also included as sample *Maple* worksheets in the **lib** directory of *Maple V Release 3* for DOS/Windows.

If you cannot find the information you seek in the online help pages, try the *Maple V Flight Manual* (by Ellis, Johnson, Lodi, and Schwalbe) or the *Maple V Quick Reference* (by Blachman and Mossinghoff), published by Brooks/Cole Publishing Company. The platform-specific documentation you received with the software (i.e., the *Getting Started* booklet or installation instructions) is an additional source of information. In many cases, this guide refers you to specific pages in other reference manuals or to help pages.

Finally, if you still have unanswered questions, contact your *Maple* technical support representatives. These people are there to help you and will do their best to answer your questions. You can reach the Technical Support Department at Brooks/Cole Publishing Company by calling (800) 214-2661, weekdays between 6:00 AM and 4:00 PM (Pacific Time); sending a fax to (408) 375-0120; or sending an email message to SUPPORT@BROOKSCOLE.COM

Interface

Many improvements have been made to *Maple V*'s user interface since its initial release. Although these changes are independent of *Maple*'s mathematical capabilities, they have been designed to increase the ease of use, the appearance, and the control of your *Maple* work.

This chapter begins with a description of *Maple V* worksheets and then outlines the platforms that support these worksheets. Improvements made in output format of expressions are described, along with a look at *Maple V*'s help facilities.

Worksheets

A *Maple* worksheet is an interface environment for combining *Maple* input, output, text, and graphics in one accessible document. The design that lets you write mathematical documents also lets you use *Maple* as a simple input/output system. For a complete description of *Maple* worksheets, refer to the interface-specific information in the *Getting Started* booklet you received with the software.

Briefly, some of the features that the worksheet environment provides are as follows:

- the ability to combine input, output, text, and graphics fields in one document
- fully editable input and text fields
- simple commands for recalculating entire worksheets
- a document transfer protocol that allows worksheets to be used easily on different platforms
- a facility that saves the state, both mathematical and visual, of your *Maple* session
- worksheet windows and two- and three-dimensional graphics windows that have buttons, menu items, and shortcut keys, allowing you, for example, to print the contents directly to a printer

Typographic Output

R3

- the ability to save entire worksheets in PostScript format for presentation purposes. In addition, a new driver has been added to *Release 3* that allows you to export a copy of a worksheet, including the mathematical output, in LaTeX format. The mathematical expressions in this document were typeset using this new feature.

Typographic Output

The *typographic* output style provides standard mathematical notation, much like that produced by sophisticated mathematical typesetting programs. For example, exponents are printed as peroperly placed superscripts, integral and summation signs are of publication quality, and many common symbols are translated to their Greek equivalents. Typographic output is available only on displays that can support bitmap graphics.

There are several different styles for mathematical output. Typographic is the default output style on platforms that support this format. The other two output styles available are *line printing* and *pretty-printing*. The style can be selected from the menu or explicitly from the command **interface(prettyprint=n)**. The example below shows how an unevaluated integral appears in all three styles, followed by some additional examples of typographic output.

```
> interface(prettyprint = 0): # for line printing
> Int(4*exp(x+alpha)/x*Pi, x=0..infinity);
Int(4*exp(x+alpha)/x*Pi,x = 0 .. infinity)

> interface(prettyprint = 1): # for pretty-printing
> int(4*exp(x+alpha)/x*Pi, x=0..infinity);
```

```
          infinity
         /
        |                 exp(x + alpha) Pi
        |         4       ----------------- dx
        |                         x
       /
        0
```

```
> interface(prettyprint = 2): # for typographic printing
> Int(4*exp(x+alpha)/x*Pi, x=0..infinity);
```

$$\int_0^\infty 4\frac{e^{(x+\alpha)}\pi}{x}\,dx$$

```
> int(1/(1+x^4),x);
```

$$\frac{1}{8}\sqrt{2}\ln\left(\frac{x^2+x\sqrt{2}+1}{x^2-x\sqrt{2}+1}\right)+\frac{1}{4}\sqrt{2}\arctan\left(x\sqrt{2}+1\right)$$
$$+\frac{1}{4}\sqrt{2}\arctan\left(x\sqrt{2}-1\right)$$

```
> array(1..3,1..3, [[exp(x), sin(x^2), 1/y], [0, gamma,
>    ln(47/3)], [delta/3, Sum(1/i,i=1..n), -2*Pi]]);
```

$$\begin{bmatrix} e^x & \sin(x^2) & \dfrac{1}{y} \\ 0 & \gamma & \ln\left(\dfrac{47}{3}\right) \\ \dfrac{1}{3}\delta & \sum_{i=1}^{n}\dfrac{1}{i} & -2\pi \end{bmatrix}$$

The text, *Maple* input and mathematical output of worksheets can be saved as a complete LaTeX file. To process this file, the two style files **maplems.sty** and **mapleenv.sty** in the **etc** subdirectory must be used. Note that this new functionality for worksheets operates independently from the existing *Maple* **latex** command. For more details see the file **maplems.tex**, which is also in the **etc** subdirectory.

Help Facilities

Maple has extensive and detailed online help. *Maple V Release 3* includes many additions to this facility, both to broaden the scope of information and to make the information easier to retrieve.

Help Topic Browser

Maple V features a hierarchical help browser. It allows users to navigate through various subjects covered by the help pages without having to use the explicit **?topic** syntax.

When the topic browser is activated, a short list of general topics (such as Graphics, Mathematics) appears. Subtopic structures can be listed in several levels of increasing detail. A brief synopsis of the current topic is printed at the bottom of the browser, to give the user a better idea of the subject of the help page.

Help Facilities

Other features include keyword search and access to multiple help pages. For more information on the help topic browser, refer to *Getting Started*.

R3

New Help Commands

Four new *Maple* commands that allow you to view specific information from a help page, have been added for *Release 3*. To retrieve specific information from the help page for **solve**, for example, use:

- **info(solve)** to display the first line of the help page, i.e., the **FUNCTION** line
- **usage(solve)** or **??solve** to display the **CALLING SEQUENCE** section of the help page
- **example(solve)** or **???solve** to display the **EXAMPLES** section of the help page, or
- **related(solve)** to display the **SEE ALSO** section of the help page.

R3

Worksheet Examples

Sample worksheets are supplied with *Maple V Release 3* in the subdirectory named **examples**. These examples are taken from applications of *Maple* and show how to use *Maple* commands for some common tasks. The following is a list of topics covered in some of these worksheets:

- Calculus
- Limits
- Plotting in Maple

There are many additional worksheets supplied in the Share Library. See the section *The Share Library* on page 28 for more details.

Mathematics

Improvements to mathematical functionality comprise most of the changes in *Maple V*. This chapter first focuses on changes in the symbolic and numeric capabilities of *Maple*. After that, some changes made to *Maple*'s ability to work in the field of complex variables are discussed. Finally, packages of library code are outlined.

Symbolics

The assume Facility

An important addition to *Maple V* is the **assume** facility. With **assume**, it is now possible to declare properties of symbolic variables without actually specifying numerical values. For example, the variable *a* can be assumed greater than zero, *b* an integer, or *c* less than *d*. You can question the properties of any variable with **is** and **isgiven**.

The uses of the **assume** facility are numerous. Many elementary problems, including some integrals and limits, cannot be solved unless assumptions are made about the values of certain variables. Previously, *Maple* returned these problems unevaluated, following its policy of not making assumptions for you. By providing assumptions before the calculation is attempted, many of these problems are solvable.

The following example demonstrates the **assume** facility and gives an example of one problem that can now be better solved. Note that the display of the variable **a** is attached with a marker to alert you that assumptions have been made.

```
> int( exp(-a*t)*t^(1/2)*ln(t), t=1..infinity );
```
$$\int_{1}^{\infty} e^{(-at)} \sqrt{t} \ln(t) \, dt$$

```
> assume(a > 0);   a;
```
$$\tilde{a}$$

```
> about(a);
```

Symbolics

```
Originally a, renamed a~:
  is assumed to be: RealRange(Open(0),infinity)
> is(a > 0);
```
$$true$$

```
> is(a < 0);
```
$$false$$

```
> int( exp(-a*t)*t^(1/2)*ln(t), t=1..infinity );
```
$$\frac{\text{MeijerG}\left(3, \frac{3}{2}, a\tilde{\,}\right)}{\sqrt{a\tilde{\,}}}$$

Fourier and Laplace Transforms

Significant additions to *Maple*'s library are the **fourier** and **invfourier** procedures, which compute symbolic Fourier and inverse Fourier transforms.

```
> readlib(fourier):
> fourier(1/(4-I*t)^(1/3),t,w);
```
$$\frac{\sqrt{3}\,\Gamma\left(\frac{2}{3}\right) e^{(-4w)} \text{Heaviside}(w)}{w^{2/3}}$$

```
> invfourier(1/(4-I*w)^(1/3),w,t);
```
$$\frac{1}{2}\frac{\sqrt{3}\,\Gamma\left(\frac{2}{3}\right) e^{(4t)} \text{Heaviside}(-t)}{\pi\,(-t)^{2/3}}$$

The **laplace** and **invlaplace** commands have been improved to handle convolutions. In the following example, the inverse Laplace transform returns an unevaluated integral.

```
> f := laplace(F(t), t, s);
```
$$f := \text{laplace}(F(t), t, s)$$

```
> invlaplace(1/(s-1)*f, s, t);
```
$$\int_0^t e^{-U} F(t - U)\, d\_U$$

```
> laplace(", t, s);
```
$$\frac{\text{laplace}(F(t), t, s)}{s - 1}$$

Integration

Many improvements have been made in both definite and indefinite integration. *Maple*'s integration facilities now handle expressions containing the following types of functions:

- elliptic integrals, are now reduced to normal form in terms of Legendre's elliptic functions

```
> f := 1/sqrt(2*t^4-3*t^2-2):
> Int(f, t=sqrt(2)..5) = int(f, t=sqrt(2)..5);
```

$$\int_{\sqrt{2}}^{5} \frac{1}{\sqrt{2t^4 - 3t^2 - 2}} dt = \frac{1}{5} \sqrt{5} \, \text{LegendreF}\left(\frac{1}{5}\sqrt{23}, \frac{1}{5}\sqrt{5}\right)$$

- algebraic functions in this example α is an alias for $\sqrt[3]{x^2 - a}$

```
> alias(alpha = RootOf(y^3-x^2-a, y)):
> f  := (4*alpha^2*x^3+(5*x^4+3*x^2*a)*alpha-3*x^2-3*a)
>          /(x^2+a)/x^2:
> int(f, x):
```

$$-3\frac{x-1}{x} + 3x\alpha + 3\alpha^2$$

- Heaviside and Dirac functions

```
> int(g(x)*Heaviside(x),x=-3..4);
```

$$\int_0^4 g(x)dx$$

- some Bessel functions

```
> int(x^9*BesselY(3,2*x^2),x);
```

$$2x^8 \, \text{BesselY}(4, 4x^2)$$

- some classes of functions whose integrals are expressed in terms of special functions

```
> g := x^(-1/3)*ln(x)*(1-x)^(1/2):
> Int(g, x=0..1) = simplify(int(g, x=0..1));
```

Symbolics

$$\int_0^1 \frac{\ln(x)\sqrt{1-x}}{x^{1/3}}\,dx =$$

$$\frac{3}{98} \frac{\left(-42\gamma + 7\pi\sqrt{3} - 63\ln(3) - 288 - 42\Psi\left(\frac{1}{6}\right)\right)\Gamma\left(\frac{2}{3}\right)\Gamma\left(\frac{5}{6}\right)}{\sqrt{\pi}}$$

```
> alias(J=BesselJ, F=hypergeom):
> assume(v>0):   assume(w>0):   assume(s>0):
> int( exp(-t^s)*t^w*J(v,t^(s/2)) , t = 0..infinity);
```

$$\frac{\Gamma\left(\frac{1}{2}\frac{2\tilde{w}+2+\tilde{v}\tilde{s}}{\tilde{s}}\right) F\left(\left[\frac{1}{2}\frac{2\tilde{w}+2+\tilde{v}\tilde{s}}{\tilde{s}}\right],[1+\tilde{v}],\frac{-1}{4}\right)}{\tilde{s}\,\Gamma(\tilde{v})\,\tilde{v}\,2^{\tilde{v}}}$$

R3
- Hermite reduction for integrals involving exponential, logarithmic, sine or cosine term. For example

```
> f := exp(-z)/(z^2*(z^2+6*z+6)^2):
> Int(f,z) = int(f,z);
```

$$\int \frac{e^{(-z)}}{z^2(z^2+6z+6)^2}\,dz = -\frac{1}{12}\frac{e^{(-z)}(2+5z+z^2)}{z(z^2+6z+6)} + \frac{1}{12}\mathrm{Ei}(1,z)$$

- improved handling of definite integrals across discontinuities.

```
> int(1/(5+3*cos(x)), x=0..13*Pi/2);
```

$$\frac{1}{2}\arctan\left(\frac{1}{2}\right) + \frac{3}{2}\pi$$

```
> int(1/x^2,x=-1..1);
```

$$\infty$$

Unevaluated Functions

R3
Use of unevaluated functions—i.e., **Int**, **Sum**, **Product**, and **Limit**— has been improved in *Release 3* to include transformations applied by **expand** and **simplify**.

```
> f := Int(sqrt(a)/Pi*exp(-u*x^2)*x^2, x):
> f = expand(f);
```

$$\int \frac{\sqrt{a}\,e^{(-ux^2)}x^2}{\pi}\,dx = \frac{\sqrt{a}\int\frac{x^2}{e^{(ux^2)}}\,dx}{\pi}$$

```
> g := Sum(a[i]*i/2^n, i=1..n):
> g = simplify(g);
```

$$\sum_{i=1}^{n} \frac{a_i \, i}{2^n} = 2^{(-n)} \left(\sum_{i=1}^{n} a_i \, i \right)$$

Differential Equations

Many improvements and additions have been made to the way *Maple* produces symbolic solutions of differential equations in *Release 3*.

- Two new algorithms for solving linear differential equations have been incorporated into *Maple*. The first algorithm either finds a solution in the domain of coefficients of the equation or provides a proof that no such solution exists. The second algorithm is similar, except that it looks for solutions in exponential extensions of the domain of the coefficients.

```
> ode := (-32*x^5+16*x^4-40*x^3+20*x^2+6*x+3)*y(x)+
>        (48*x^5+44*x^3-15*x-3)*diff(y(x),x)+
>        (-16*x^5-24*x^4-26*x^2+9*x)*diff(y(x),x$2)+
>        (8*x^4-4*x^3+6*x^2)*diff(y(x),x$3);
```

$$ode := (-32x^5 + 16x^4 - 40x^3 + 20x^2 + 6x + 3)y(x)$$
$$+ (48x^5 + 44x^3 - 15x - 3)\left(\frac{\partial}{\partial x} y(x)\right)$$
$$+ (-16x^5 - 24x^4 - 26x^2 + 9x)\left(\frac{\partial^2}{\partial x^2} y(x)\right)$$
$$+ (8x^4 - 4x^3 + 6x^2)\left(\frac{\partial^3}{\partial x^3} y(x)\right)$$

```
> dsolve(ode, y(x));
```

$$y(x) = \_C1 \, e^x + \_C2 \, e^x \, \text{Ei}(1, -x) + \_C3 \, e^x \int e^{(x^2)} \sqrt{x} \, dx$$

- In linear ordinary differential equations, **dsolve** returns more solutions, by taking advantage of the **RootOf** function.

Symbolics

```
> de := diff(y(x),x$6) + 4*diff(y(x),x$4) + 4*diff(y(x),x$3)
>     + 4*diff(y(x),x$2) + 8*diff(y(x),x);
```

$$de := \left(\frac{\partial^6}{\partial x^6} y(x)\right) + 4\left(\frac{\partial^4}{\partial x^4} y(x)\right) + 4\left(\frac{\partial^3}{\partial x^3} y(x)\right) + 4\%1 + 8\left(\frac{\partial}{\partial x} y(x)\right)$$

$$\%1 := \frac{\partial^2}{\partial x^2} y(x)$$

```
> dsolve(de, y(x));
```

$$y(x) = \_C2 + \_C3 \cos\left(\sqrt{2}x\right) + \_C4 \sin\left(\sqrt{2}x\right) + \left(\sum_{\_R=\%1} \_C1\_R e^{(\_Rx)}\right)$$

$$\%1 := \operatorname{RootOf}(\_Z^3 + 2\_Z + 4)$$

R3

- Output forms can now be specified by optional arguments. For example, **output=basis** results in the basis of the solution space of a linear ODE, along with any particular solutions.

```
> de := 2*diff(y(x),x$2) + 3*diff(y(x),x) +y(x) - exp(-3*x);
```

$$de := 2\left(\frac{\partial^2}{\partial x^2} y(x)\right) + 3\left(\frac{\partial}{\partial x} y(x)\right) + y(x) - e^{(-3x)}$$

```
> dsolve(de, y(x), output=basis);
```

$$\left[[e^{(-x)}, e^{(-1/2x)}], \frac{1}{10} e^{(-3x)}\right]$$

R3

- When using the **series** or **laplace** options to **dsolve**, you can now provide *nonzero*, algebraic, initial conditions, such as **y(1)=a, D(y)(3)=b**

R3

- If the complete solution to a linear differential equation cannot be found in closed form, a partial solution containing **DESol** functions is returned.

```
> de := diff(y(x),x$3) + (2*x+2)*diff(y(x),x$2) +
>     (4*x+4-1/x)*diff(y(x),x) + 2*y(x);
```

$$de := \left(\frac{\partial^3}{\partial x^3} y(x)\right) + (2x+2)\left(\frac{\partial^2}{\partial x^2} y(x)\right) + \left(4x+4-\frac{1}{x}\right)\left(\frac{\partial}{\partial x} y(x)\right) + 2y(x)$$

```
> dsolve(de, y(x));
```

$$y(x) = \_C1\, e^{(-x^2)} + e^{(-x^2)} \int \text{DESol}\left(\left\{x\left(\frac{\partial^2}{\partial x^2}\_Y(x)\right)\right.\right.$$
$$+ (-4x^2 + 2x)\left(\frac{\partial}{\partial x}\_Y(x)\right) + (-4x^2 - 2x - 1 + 4x^3)\_Y(x)\right\},$$
$$\left.\{\_Y(x)\}\right) dx$$

DESol represents the solution of a differential equation without actually computing the solution, thereby allowing you to manipulate the solution symbolically before attempting further computation. The available commands for operation on **DESol** functions include **diff**, **int**, **series**, and **evalf**.

For more information and examples of all the above improvements, see the *Differential Equations* section of **?updates,v5.3**

The minimize command

The **minimize** command now supports an optional third parameter, which specifies the real range over which a minimal value is found. The default range is the open range $(-\infty, \infty)$. Passing **infinite** as the third parameter specifies the closed range $[-\infty, \infty]$. Minimization is also supported over finite intervals.

```
> f := x^3 - 5*x^2 + 4;
```
$$f := x^3 - 5x^2 + 4$$

```
> minimize(f, x, infinite);
```
$$-\infty$$

```
> minimize(f, x, {x=0..5});
```
$$\frac{-392}{27}$$

Simplification of Radicals

Previous to *Maple V Release 3*, *Maple* performed simplifications and transformation of radicals that were not valid in *all* cases. In *Release 3*, the following simplifications and transformations are made given the appropriate restrictions, but not otherwise.

Symbolics

- $(xy)^{n/d} \to x^{n/d} \, y^{n/d}$
- $(x^r)^{n/d} \to x^{rn/d}$

Transformations that are *provable* are still made automatically, including:

- $n^{a/b} \to n^{q+r/b} \to n^q \, n^{r/b}$, where $n, a, b \in I$
- $(n/d)^r \to \frac{n^r}{d^r}$, where $n, d \in I$ and $r \in Q$
- $(x^r)^{n/d} \to x^{rn/d}$, where $n, d \in I$ and $r \in Q$

R3 Similarly, transformations indicated by applying the commands **sqrt**, **simplify**, **combine**, and **expand** to square roots and radicals are no longer automatically made. For example,
```
> sqrt( Pi^2 * x^2 );
```
$$\pi \sqrt{x^2}$$

R3 Transformations can still be forced by using the **symbolic** option to the **sqrt** command, or by using the **assume** command, to declare restrictions on the values of variables. An example of the former command is
```
> sqrt( Pi^2 * x^2, symbolic );
```
$$\pi x$$

and of the latter command is
```
> assume( x>0 );
> sqrt( Pi^2 * x^2 );
```
$$\pi \tilde{x}$$

R3 The **simplify** command for expressions containing square roots and radicals has been improved. The principal transformation made is

- $(a^n \, b)^{m/d} \to (a^{q+r/d} \, b)^{m/d} \to a^{mq} \, (a^{r/d} \, b)^{m/d}$, where $a > 0$.

Similar options are available with the **combine** command, where the principal transformation made is

- $x^{m/d} y^{n/d} \to (x^m y^n)^{1/d}$ iff signum(x) > 0 or signum(y) > 0.

Both commands are illustrated in this example.
```
> f := a^(1/3)*(b^3*c^2)^(1/3);
```
$$f := a^{1/3} (b^3 c^2)^{1/3}$$

```
> simplify(f, radical, symbolic);
```
$$a^{1/3} b c^{2/3}$$

```
> combine(", radical, symbolic);
```
$$b (a c^2)^{1/3}$$

R3 Denesting of radicals has been improved in the **sqrt** and **simplify** commands. Additionally, an even more powerful denesting tool, **radnormal**, has been provided.
```
> f := sqrt(2*(3-sqrt(2)-sqrt(3)+sqrt(6)));
```
$$f := \sqrt{6 - 2\sqrt{2} - 2\sqrt{3} + 2\sqrt{6}}$$

```
> readlib(radnormal):
> radnormal(f);
```
$$-1 + \sqrt{3} + \frac{1}{3}\sqrt{3}\sqrt{6}$$

R3 The command **rationalize** rationalizes expressions containing radicals.
```
> f := 1/(x^(4/3)+y^(3/2));
```
$$f := \frac{1}{x^{4/3} + y^{3/2}}$$

```
> readlib(rationalize):
> rationalize(f);
```
$$-\frac{(y^{3/2} - x^{4/3})(x^{16/3} + x^{8/3} y^3 + y^6)}{x^8 - y^9}$$

Symbolics

Algebraic Numbers, Functions, and Fields

Several enhancements have been made to *Maple*'s handling of algebraic numbers, functions, and fields.

- *Maple* can factor multivariate polynomials over any algebraic number field. For example, in the following eigenvalue problem the characteristic polynomial is factored over the algebraic number field $Q(\sqrt{2})$.

```
> with(linalg):
> T := toeplitz( [sqrt(2), x, 2*omega] );
```

$$T := \begin{bmatrix} \sqrt{2} & x & 2\omega \\ x & \sqrt{2} & x \\ 2\omega & x & \sqrt{2} \end{bmatrix}$$

```
> charpoly(T, lambda):
> factor(");
```

$$-\left(-\lambda^2 + 2\lambda\sqrt{2} + 2\omega\lambda - 2 - 2\omega\sqrt{2} + 2x^2\right)\left(-\sqrt{2} + 2\omega + \lambda\right)$$

```
> eigenvals(T);
```

$$\sqrt{2} + \omega - \sqrt{\omega^2 + 2x^2},\ \sqrt{2} + \omega + \sqrt{\omega^2 + 2x^2},\ -2\omega + \sqrt{2}$$

- **evala** has been extended to algebraic functions.

```
> alias(sqrtx=RootOf(z^2-x,z)):
> evala(Factor(x*y^2-1,sqrtx));
```

$$x\left(y + \frac{sqrtx}{x}\right)\left(-\frac{sqrtx}{x} + y\right)$$

- *Maple* can compute with and factor multivariate polynomials over general finite fields $GF(p^k)$ as well as the integers modulo p a prime. *Maple* represents elements of $GF(p^k)$ as polynomials in α, where α is a **RootOf** an irreducible polynomial. Here is an example of a univariate factorization over G(9)

```
> alias( alpha=RootOf(x^2+2*x+2) ):
> f := 2*alpha*x^3+(2+2*alpha)*x^2+(alpha+2)*x+1;
```

$$f := 2\alpha x^3 + (2 + 2\alpha)x^2 + (\alpha + 2)x + 1$$

```
> Factor(f) mod 3;
```

$$2\alpha(x + \alpha)(x + \alpha + 2)(x + 1 + 2\alpha)$$

Mathematics

R3
- Here is an example of multivariate factorization
```
> f := x^3+alpha*x^2*y+alpha^2*x*y^2+alpha^3*y^3;
```
$$f := x^3 + \alpha x^2 y + \alpha^2 x y^2 + \alpha^3 y^3$$

```
> Factor(f) mod 3;
```
$$(y + \alpha x)(y + 2\alpha x)(y + (\alpha + 2)x)$$

Additionally, integration of algebraic functions and multivariate polynomial factorization over algebraic number fields are now available.

R3
In *Release 3*, further refinement has been made to *Maple*'s facilities for solving algebraic equations. Improvements have been made in the areas of selection of roots, solving large systems and systems of trigonometric equations, and equations containing **RootOf**s. See **?updates,v5.3** for examples of all these improvements.

R3
The **RootOf** command has been extended with an optional third parameter, which allows identification of an individual root closest to a given numeric value.
```
> RootOf(x^4-2, x, 1.2);
```
$$\text{RootOf}(\_Z^4 - 2, 1.2)$$

```
> evalf(");
```
$$1.189207115$$

Additional Functionality

- Symbolic simplifications are now available for **min, max, round, ceil, floor, Re, Im, Heaviside,** and **Dirac**.

R3
- The **convert** command has a new **binomial** option for converting factorials and **GAMMA**s into binomials.

R3
- The **combine(** *expr*, **ln)** command takes, as a new optional parameter, a *Maple* type name specifying what type of logarithms you want to combine.

- Trigonometric formulas are implemented for multiples of $\frac{\pi}{8}$, $\frac{\pi}{10}$, and $\frac{\pi}{12}$.

- Derivatives are now computed for **signum, csgn, trunc, frac, round, floor,** and **ceil**.

Numerics

- The composition operators, @ and @@, are now handled by **expand**.

- The shifted sine integral, **Ssi**, the hyperbolic sine integral, **Shi**, and the hyperbolic cosine integral, **Chi**, have been added.

R3
- The command **piecewise** now allows you to define piecewise-continuous functions. As well, global smoothness can be specified at the joints. See the online help page for **piecewise** for more information.

R3
- The value of **signum(0)** can now be defined by the user. For an explanation of the importance of this change, see **?updates,v5.3**.

Numerics

Maple's numerical functionality has been continually improved and updated. Enhancements in speed, accuracy, and breadth of functionality have been made.

dsolve(,numeric)

R3
The operation of the differential equation solver, **dsolve**, has been improved for *Release 3*. When **dsolve** is called with the **numeric** option, a *Maple* procedure is returned. This procedure allows the user to calculate individual values of the solution.

```
> de := diff(y(t),t$3) - 2*diff(y(t),t$2) + 2*y(t);
```

$$de := \left(\frac{\partial^3}{\partial t^3} y(t)\right) - 2\left(\frac{\partial^2}{\partial t^2} y(t)\right) + 2y(t)$$

```
> initcond := y(0)=1, D(y)(0)=1, (D@@2)(y)(0)=1:
> G := dsolve({de,initcond}, y(t), numeric);
G := proc(rkf45'x) ... end

> G(0.2);
```

$$\left[t = .2, y(t) = 1.219849548122092, \frac{\partial}{\partial t} y(t) = 1.196899483561325, \frac{\partial^2}{\partial t^2} y(t) = .9511440965846975\right]$$

As you can see, invoking the procedure returned by **dsolve** produced a **list** of equations. This format allows for simple substitutions. Such substitutions are used in the **plots** package procedure **odeplot**, which provides a convenient method of plotting solution curves of differential equations in both two and three dimensions.

```
> plots[odeplot](G, [diff(y(t),t),diff(y(t),t$2)], -4..4,
>                -20..5, numpoints=50);
```

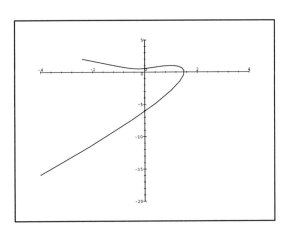

R3 A further optional parameter can be added to the call to **dsolve** to change the output format from a single procedure that produces a list to a list of procedures producing single outputs. These procedures can then be treated much as any other *Maple* function and plotted with the standard plotting routines. In the following example, $y(t)$ and its first and second derivatives are plotted against each other as a curve in three-dimensional space.

```
> H := dsolve({de,initcond}, y(t), numeric, output=listprocedure);
  H := [t = proc(t) ... end, y(t) = proc(t) ... end,
```

$$\frac{d}{dt} y(t) = \text{proc}(t) \ldots \text{end}, \quad \frac{d^2}{dt^2} y(t) = \text{proc}(t) \ldots \text{end}]$$

```
> yt := rhs(H[2]): dyt := rhs(H[3]): ddyt := rhs(H[4]):
> plots[spacecurve]([yt,dyt,ddyt,0..1.5], labels=[y,dy,ddy],
>                   axes=frame);
```

Complex Variables

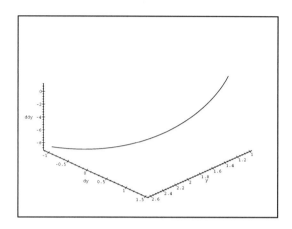

Additional Functionality

- The new implementation of **fsolve** provides more effective and accurate numerical solution of expressions over the real line or complex plane.

- The logarithmic integral, **Li(x)**, which provides an approximation to the number of primes less than **x**, has been added.

- The numerical evaluation of infinite sums, products, and limits has been improved by the addition of Levin's *u* transform.

- In the **linalg** package, matrices of floating-point and complex floating-point elements are now handled by the procedures **colspace**, **det**, **eigenvals**, **eigenvects**, **gausselim**, **gaussjord**, **inverse**, and **rowspace**.

Complex Variables

Many improvements have been made to the way *Maple* handles variables in the complex plane.

Mathematics

Complex Data Type

New data types have been added to handle complex variables. The command **type(expr,complex(t))** tests if an expression, **expr**, is of the form **a + b*I** where **a** and **b** are of type **t**. Typical cases are as follows:

- **complex(integer)**, i.e., Z[i], the Gaussian integers
- **complex(rational)**, i.e., Q[i]
- **complex(float)**, i.e., a complex floating-point number
- **complex(numeric)**, which is any of these.

Automatic Complex Arithmetic

Complex numerical arithmetic is now automatic; that is, addition, subtraction, multiplication, division, and exponentiation of complex integers, complex rationals, and complex floats is done automatically. Also, **evalf** now works over the complex floats. Consequently, many functions with complex numbers work better and faster than in previous versions.

```
> (2/3+4/5*I)^2;
```
$$-\frac{44}{225} + \frac{16}{15}I$$

More Functions Supported

The following is a list of the *Maple* functions with improved handling of complex variables.

- **exp** and **ln**
```
> exp(1.234+5.678*I);
```
$$2.824884809 - 1.954188170\,I$$

- trigonometric functions and inverse trigonometric functions
- exponential/**sin/cos/ sinh/cosh/log** integrals

Packages

- **GAMMA**, **Beta**, and **Psi** (and its derivatives)
```
> Beta(1.2+3.4*I,-2.1+5.7*I);
                    .6600944470 − 1.126821143 I
```

- **Zeta** (and its derivatives)
- **erf**, **erfc** (including iterated integrals)
- **W** (all branches)

Additional Functionality

- The routines for **Re(z)**, **Im(z)**, **argument(z)**, and **conjugate(z)** have been improved; and the routine **csgn(z)**, which computes the complex sign of **z**, has been added.

Packages

Much of *Maple*'s mathematical functionality is contained in its library of specialized packages. Several new packages have been added, and other existing packages have been thoroughly updated. For more information on what packages are available, see the help page for **?packages**. Each package has a related help page that details the individual commands contained in the package.

Packages New to Release 2

- A **networks** package provides an environment for constructing, drawing, and analyzing properties of combinatorial networks. Networks are represented as graphs with multiple edges, possibly directed, and loops. Edge and vertex weights can be arbitrary *Maple* expressions.

- The numerical approximation package, **numapprox**, has been added. The main function in this package is **minimax**, which uses the Remez algorithm to compute the best minimax polynomial of degree $\leq n$, or rational function with numerator of degree $\leq m$, and denominator of degree $\leq n$ to the function f on the interval [a,b].

- The **padic** package has been added to deal with the arithmetic of p-adic numbers.

- The **GaussInt** package has been added for manipulation of Gaussian Integers.

- The **genfunc** package has been added for manipulation of rational generating functions.

Mathematics

Packages Improved for Release 2

- The number theory package, **numtheory**, has many new routines. The **cfrac** routine now computes continued fraction expansions for real numbers, polynomials, series, and formulas. Several forms are available, including **simple**, **regular**, and **simregular**. The utility routines **nthconver**, **nthnumer**, and **nthdenom** return the nth convergent, numerator, and denominator of the continued fraction expansion.

- The linear algebra package, **linalg**, has many new procedures, including **blockmatrix**, **companion**, **entermatrix**, **pffge**, **randvector**, **rcf**, and **Wronskian**.

- The **liesymm** package has been extended to include several new facilities and routines for the recovery, manipulation, and conversion of systems of partial differential equations.

Packages New to Release 3

The **stats** package has been completely redesigned and rewritten. The functionality in **stats** is broken down into *subpackages*, which are listed when **stats** is loaded.

```
> with(stats);
```
$$[\,describe, fit, importdata, random, statevalf, statplots, transform\,]$$

The basic data structure of the **stats** package is a *statistical list*. In its simplest form, a statistical list is an ordinary list of values; but it can also contain *ranges* and *weighted* values. The following are valid statistical lists.

```
> sl1 := [1, 3, 8, 7, 8, 9, 3, 10, 5, 6, 8];
```
$$sl1 := [\,1, 3, 8, 7, 8, 9, 3, 10, 5, 6, 8\,]$$

```
> sl2 := [1, 3..5, 7, 6..9, Weight(2, 5), 6, Weight(7..10, 2)];
```
$$sl2 := [\,1, 3..5, 7, 6..9, \text{Weight}(\,2, 5\,), 6, \text{Weight}(\,7..10, 2\,)\,]$$

```
> sl3 := [Weight(1..3, 5), Weight(4..6, 9), Weight(7..10, 3)];
```
$$sl3 := [\,\text{Weight}(\,1..3, 5\,), \text{Weight}(\,4..6, 9\,), \text{Weight}(\,7..10, 3\,)\,]$$

Commands in the subpackages can then be used to investigate the data.

```
> describe[mean](sl1);
```
$$\frac{68}{11}$$

Packages

```
> describe[median](sl1);
```
$$7$$

```
> transform[split[2]](sl3);
```
$$\left[\left[\text{Weight}(1..3, 5), \text{Weight}\left(4..6, \frac{7}{2}\right)\right],\right.$$
$$\left.\left[\text{Weight}\left(4..6, \frac{11}{2}\right), \text{Weight}(7..10, 3)\right]\right]$$

There are also extensive plotting commands provided in the **statplots** subpackage.

```
> slrand1 := [random[normald](100)]:
> ranges := [-5..-2, -2..-1, -1..0, 0..1, 1..2, 2..5]:
> group1 := transform[tallyinto](slrand1, ranges);
```
$$group1 := [\text{Weight}(-1..0, 27), \text{Weight}(0..1, 44), \text{Weight}(2..5, 3),$$
$$\text{Weight}(1..2, 12), \text{Weight}(-2..-1, 12), \text{Weight}(-5..-2, 2)]$$

```
> statplots[histogram](group1);
```

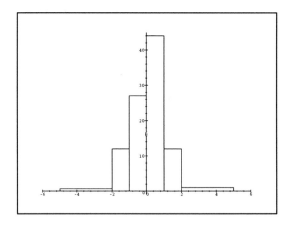

```
> slrand2 := [random[normald](100)]:
> slrand3 := [random[normald](100)]:
> statplots[scatter2d](slrand2, slrand3);
```

Mathematics

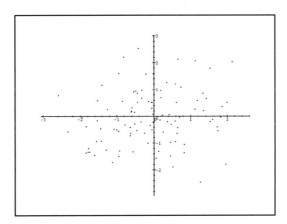

For more information on the commands available in the **stats** package, see any one of the numerous help pages. Also, the help page **?stats[updates]** has details of the features of the new **stats** package, including instructions on how to update your *Release 2* code to take advantage of the *Release 3* changes.

Packages Improved for Release 3

R3
- The number theory package, **numtheory,** has one new routine. **factorEQ** factors integers in a Euclidean domain. The commands **isprime** and **safeprime** have been updated to perform one strong pseudo-primality test and one Lucas test. The speed of **jacobi** has been improved by a factor of 15.

R3
- The linear algebra package, **linalg**, has improved commands in two areas. First, **curl, diverge, grad**, and **laplacian** have been extended to work in spherical and cylindrical coordinate systems. Second, **hermite, ihermite, smith**, and **ismith** now optionally return multiplier matrices. For more information, see the individual help pages for these commands.

R3
- The **GaussInt** package has four new commands. **GIsmith** and **GIhermite** compute the appropriate normal forms over $Z[i]$. **GIchrem** implements the Chinese

Remainder Theorem over $Z[i]$. **Glnodiv** computes the number of nonassociated divisors.

R3
- The the parameter ordering of **grobner[gsolve]** has been changed to be more consistent with other *Maple* solvers. See the online help page for details.

The Share Library

R3
The share library has been updated for *Release 3*. The share library is a repository of *Maple* code submitted by users of *Maple*. Available as part of the share library are

- more than 25 packages of *Maple* code
- more than 50 individual *Maple* procedures
- more than 70 *Maple* worksheets
- more than 25 articles and papers concerning *Maple*.

Most of these files are not included with the Student Edition of *Maple V Release 3*, but they can be obtained through electronic mail. For more information, see the help file **?share** or chapter 5, *Electronic Resources*, in *Maple V Quick Reference* by Blachman and Mossinghoff.

A sampling of the new topics covered includes

- permutation groups
- C and Fortran language code generation
- parametric surfaces in space
- Puiseux series expansion
- reliability polynomial of a network
- implementation of Gröbner basis algorithms
- matrix Pade function approximation
- Ritt-Wu's characteristic sets

Mathematics

- Polynomial and rational control systems
- interval arithmetic
- matrix normal forms
- generating functions, recurrences, and ODEs
- finite Coxeter groups
- formal power series
- educational worksheets in science, engineering and calculus.

Also, the mechanism for accessing the share library has been greatly simplified. Consult your *Getting Started* or the **?updates,v5.3** help page for more information on the contents of the share library.

Graphics

Plotting of mathematical expressions in both two and three dimensions has improved, and the respective plotting facilities are now closer in functionality. Animation in both two and three dimensions extends *Maple*'s visualization applications. A larger number of output formats and devices are also supported.

Contour Plots

Maple provides contour plotting. Most surfaces can also be rendered as a set of contours. The following is an example of a contour plot for the mathematical representation of a vibrating drum head. The number of contours plotted can be controlled with the **contours** option.

```
> k := 11.61984117:
>
> plot3d([r*cos(t),r*sin(t),BesselJ(2,k*r)*cos(2*t)],
>     r=0..1, t=0..2*Pi, grid=[30,60], scaling=CONSTRAINED,
>     style=CONTOUR, orientation=[90,0]);
```

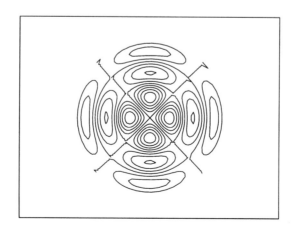

Implicit Plots

R3

Maple can now plot implicit equations in two- and three-dimensions with the **implicitplot** procedure in the **plots** package. In *Release 3*, the speed of **implicitplot** has been greatly improved.

```
> with(plots):
> f := (x-3)^2/25 + (y+1)^2/9 = 1;
```

$$f := \frac{1}{25}(x-3)^2 + \frac{1}{9}(y+1)^2 = 1$$

```
> implicitplot(f, x=-10..10, y=-5..5, scaling=CONSTRAINED,
>     axes=NORMAL);
```

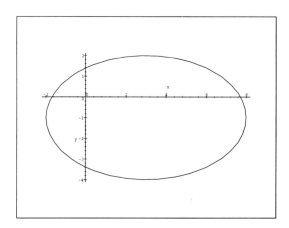

Vector Field and Gradient Vector Field Plots

The new *Maple* procedures **fieldplot** and **fieldplot3d** generate and plot a two- or three-dimensional vector field, respectively, while **gradplot** and **gradplot3d** generate and plot a two- or three-dimensional gradient vector field.

```
> with(plots):
> gradplot(sin(x*y),x=-Pi..Pi,y=-Pi..Pi,arrows=SLIM);
```

Graphics

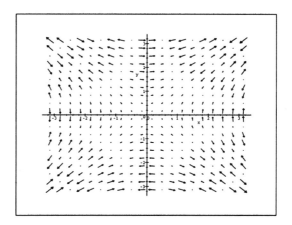

Scaling Axes

Another new feature in *Maple* is the ability to scale the axes of a plot. Log and log-log plots are now easily created with **logplot** and **loglogplot** from the **plots** package.

```
> with(plots):
> loglogplot(10^x, x=1..10);
```

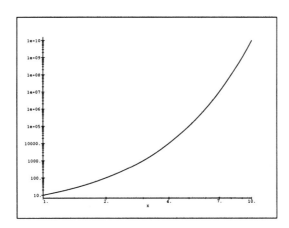

Point Style

R3

New to *Release 3* is the **symbol** option, which allows you to specify the symbol used to represent a point in two- and three-dimensional plots. Available choices include:

BOX	□
CROSS	×
CIRCLE	○
DIAMOND	◇
POINT	•

Line Style

R3

The style of line used in a plot can be controlled with the **linestyle** option. Available line styles differ from system to system but are indicated with an integer value between 1 and 7. The following plot demonstrates some of the options detailed in the previous three sections.

```
> A := plot(sin(x)/x, x=-10..10, title=`sin(x)/x`, thickness=2):
> estimate := convert(series(sin(x)/x, x=0, 20), polynom):
> B := plot(estimate, x=-10..10, style=POINT, symbol=CIRCLE):
> plots[display]({A,B});
```

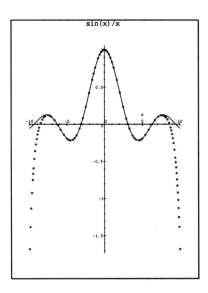

Graphics

Line Thickness

R3   The thickness of lines drawn in a plot can be controlled with the **thickness** option. The thickness can be set to a nonnegative integer value such as 0, 1, 2, or 3. A thickness of 0 refers to the default thickness for your output device.

Plot Labeling

Parameters have been added for control of plot labeling. The number of tickmarks on each axis can be controlled with the **xtickmarks** and **ytickmarks** commands. Text can be overlaid on a plot using **textplot** with **display** or **textplot3d** with **display3d**. New to Release 3 is the ability to set the font family, size, and style for text in a plot

R3   by using the **font** option. See **?plot[options]** for further details.

Rendering

Maple's three-dimensional rendering schemes have improved significantly. A new algorithm for surface rendering has eliminated anomalies previously found in surface plots. A complete list of the rendering choices now available is as follows:

- Wireframe
- Patch
- Patch Without Grid
- Contours
- Patch and Contour
- Hidden Line (wireframe plot with hidden line removal)
- Point (point plot or scatter plot).

Lighting Models

Maple has the ability to apply lighting and shading models to three-dimensional plots. Through use of the **light** and **ambientlight** options, either directed or ambient light

Coloring

sources may be added to a plot. Multiple directional lights can also be combined for one surface.

When specifying directed light, the following values must be given: spherical coordinates phi and theta in degrees of the light direction; and red, green, blue (RGB) values each between 0 and 1.

light = [*phi_value, theta_value, red_value, green_value, blue_value*]

When specifying ambient light, no directional coordinates need be given but RGB values must be between 0 and 1.

ambientlight = [*red_value, green_value, blue_value*]

With most platforms, three or four predefined lighting models are supplied as preset menu items.

Coloring

Two-dimensional curves can now be plotted with an assigned color. Colors can be specified by **RGB** values or **hue** values, or a **color** from 25 predefined names. For more information, see the help page for **?plot[color]**.

```
> plot(sin, color = COLOR(RGB,0.2,0.3,0.5));
> plot(sin, color = COLOR(HUE,.1));
> plot(sin, color = blue);
```

Many changes have also been made to *Maple*'s three-dimensional coloring schemes. The color ranges for the **Z, XY, XYZ** options have been altered to improve presentation of the plots. Other predefined schemes include:

- **ZHUE** - rainbow colors along Z axis
- **ZGRAYSCALE** - gray scale along Z axis
- **NONE** - black and white only

Note that color can be added to any three-dimensional plots through the use of colored light sources.

Additionally, any surface can be colored using any expression in the current variables.

Graphics

For more information on this option, see the help pages for **?plot3d** and **?plot3d[colorfunc]**.

Plot Data Structure

The numerous changes to three-dimensional plotting capabilities have, of course, led to many changes in the *Maple* data structure representing these plots. *Maple* programmers who manipulate data structures are referred to the appropriate help page using **?plot3d[structure]**.

Other Improvements

- Two- and three-dimensional rendering of convex polygons has been added. This allows for the creation of three-dimensional polyhedra. See **?plots[polygons]** and **?plot[polyhedraplot]** for more details.

- The **style=SPLINE** option of the **plot** command has been removed from *Maple V*.

Animation

Two- and three-dimensional animation are available on *Maple* platforms that can support the necessary calculations and display. This feature allows you to view conventional plots as they vary with respect to an additional variable. Parameters are supplied to control the number of frames produced, as well as the speed and direction of motion. Interface controls as found on a videocassette recorder (i.e., play, stop, fast forward, and so on) are provided in an animation window.

The functions for animation are **animate** and **animate3d**. Because it is impossible to show you the full effects of animation in a printed document such as this, we urge you to try the following examples in *Maple V*. Note that the more frames that you specify, the more initial calculations must be done, the more memory you will need, and the longer you will have to wait for your animation.

Note: In the Student Edition, the value of frames * numpoints cannot exceed 5,120.

For more information on *Maple*'s animation facility, see the help pages for **?animate**, **?animate3d**, **?plot**, and **?plot3d**.

Two-Dimensional Animation

To create a two-dimensional animation, supply an expression in two unknowns and then specify ranges for both unknowns. The last unknown that you specify will be taken as the variable of animation, that is, the one that is varied to create the separate frames of the animation.

```
> with(plots):
> animate(sin(x*t), x=-10..10, t=1..2, frames=50);
> animate([u*t,t,t=1..8*Pi], u=1..4, coords=polar, frames=60);
```

Three-Dimensional Animation

To create a three-dimensional animation, supply an expression in three unknowns and then specify ranges for all the unknowns. As in two-dimensional animation, the last unknown that you specify will be taken as the variable of animation.

```
> with(plots):
> animate3d((1.3)^x * sin(u*y), x=-1..2*Pi, y=0..Pi, u=1..8,
>    coords=spherical);
> animate3d([x*u,t-u,x*cos(t*u)], x=1..3, t=1..4, u=2..4);
```

Output Formats and Devices Supported

Additions and improvements have been made to Maple's graphics to support more output formats, printers and plotters. The following formats and devices are supported on *all* Maple V platforms.

Encapsulated PostScript, **ps**, and *color PostScript*, **cps**, output formats are available for both two- and three-dimensional plots.

Three-dimensional plot support has been added for *Tektronix*, **tek**, devices. Other cross-platform formats and devices supported include:

- character plots

- **i300** - Imagen 300 printers

- **ln03** - DEC LN03 printers

- **pic** - troff pic language

- **regis** - DEC REGIS terminals

Graphics

- **unix** - UNIX plot facility
- **vt100** - vt100 terminals (animation not supported)
- **hplj** - HP Laserjet printers
- **hpgl** - HP plotters

System and Language

Improvements to the structure of *Maple*'s programming language and the way *Maple* interacts with host operating systems are important changes in *Maple V Release 3*. The handling of library files and other external input/output files make up a large proportion of these changes. Enhancements to general methods of error handling and debugging are also discussed. Finally, changes to *Maple*'s programming language are given.

Libraries and I/O

Multiple Library Access

Maple V Release 3 supports multiple user libraries. The ability to specify multiple libraries makes code development a much more easily controlled task. Multiple libraries are specified with a statement such as:

libname := '/usr/maple/lib', '/u/myself/ mydirectory';

Code can be segmented into logical packages and *Maple*'s standard library can be easily and safely augmented or overridden. When a previously undefined *Maple* procedure is used in a *Maple* session, each of the specified libraries is searched, in order, to find the necessary information. Specifying a library before the standard library (represented in the above example by **/usr/maple/lib**) would result in a library override.

For more information on multiple libraries, see the help page **?libname**.

Maple Library Archiver

On all *Maple* platforms, thousands of *Maple* library files have been archived into one large file. A new tool, **march**, is included with all platforms, allowing you to add or update **.m** files in a *Maple* library archive. For more information, consult your platform specific documentation.

Other options allow you to investigate the structure of a current archive, insert files in

Libraries and I/O

a new or existing archive, extract files from an archive, and compact memory storage in an archive. For more information, see the related help page on your system.

Formatted Input and Output

Formatted I/O in *Maple V* allows you to read data from files created by other applications and to store *Maple* results in formats readable by other applications.

The procedures provided are **printf, sscanf, readline, readdata,** and **parse**.

The first two resemble their namesakes from the C programming language. For reading in a line from a file, use **readline**; to read data arranged in columns into a list of lists, use **readdata**; to pick individual items out of that line, use **sscanf**; to print information back out, use **printf**.

Consider a data file, *mydata*, four lines long with the following information:

```
12  4  3 -5
 9  4 -2 91
32 96  0  2
-1  5  4 96
```

The following *Maple* code will read these numbers into a list of lists; create a matrix, *A*, from that list; calculate the transpose, *At*, of *A*; and then output the results in a similar format to the input file.

```
> readlib(readdata):
> readdata(mydata, integer, 4);
> A := convert(",matrix):
> At := linalg[transpose](A):
>
> for i from 1 to 4 do
>     printf('%d %d %d %d\n', At[i,1],At[i,2],At[i,3],At[i,4]);
> od;
```

The resulting file would look like

```
12  9 32 -1
 4  4 96  5
 3 -2  0  4
-5 91  2 96
```

System and Language

The **parse** command parses a string as a *Maple* statement. Statements can be either fully evaluated immediately or have their evaluation delayed, depending on whether the **statement** option is specified.

```
> parse('sin(3.0) + cos(3.0);');
```
$$\sin(3.0) + \cos(3.0)$$

```
> ";
```
$$-.8488724885$$

```
> '(2.0 + exp(2.0))^2.0;': # a string
> parse(", statement);
```
$$88.15437443$$

.m File Format

The **.m** format that *Maple* uses to save binary files has changed in *Release 3*. To re-create your **.m** files under *Maple V Release 3*, reload your source code and save it again with a **.m** extension.

R3 The **m2src** tool has been added to convert *Release 2* **.m** (*Maple*'s internal format) files to *Release 2* source files. The **updtsrc** tool, which converts *Release 2* source files to *Release 3* source files, has been added. This latter tool inserts appropriate **global** statements, so that variables in procedures that were global in *Release 2* remain global in *Release 3*. See the section *Global Variables* on page 45 for more information.

Clearing the Session

Maple V has a command, **restart**, that allows you to clear all the assignments made during a session, basically doing an internal restart of the system. This command clears all assigned names, unloads any packages or functions you may have loaded, and unloads any procedures you may have defined.

Error Handling and Debugging

Several enhancements have been made to *Maple*'s facilities for error handling and debugging of code. Although they have the greatest effect on users creating their own *Maple* procedures, these tools are also valuable when you want detailed information on how *Maple* is operating.

Error Handling and Debugging

Automatic Type Checking

Automatic type checking allows you to write *Maple* procedures that will operate only on parameters of specified type(s). Although you often want procedures to work on a wide spectrum of types, there are cases when only one or two types are desired. In this case, automatic type checking is invaluable.

Standardized error messages are produced when a procedure, either written by the user or within *Maple*'s standard library, is used and fails automatic type checking. The following example shows a procedure, **appendtolist**, that takes a **list** as its first parameter and an **integer** as its second parameter, and appends the **integer** to the beginning of the **list** to form a new **list**.

```
> appendtolist := proc(alist:list, n:integer)
>    [n, op(alist)];
> end:
```

Here is what happens when erroneous parameters are passed to **appendtolist**.

```
> appendtolist([1,2,3,4],5.6);
Error, appendtolist expects its 2nd argument, n,
to be of type integer, but received 5.6

> appendtolist([1,2,3]);
Error, (in appendtolist) appendtolist uses a 2nd argument, n
(of type integer), which is missing
```

errorbreak

The **errorbreak** command provides user control over how *Maple* behaves when encountering errors while reading a file. There are three choices that can be set in the **errorbreak** option of the **interface** command.

With **interface(errorbreak=0)**, there is no error breaking. The entire input file is read in and all the appropriate error messages are displayed. This is the way *Maple* has performed in the past.

With **interface(errorbreak=1)**, input stops after the first *syntax error* is encountered, thus avoiding subsequent syntax error messages that may only be consequences of the first syntax error. This is now the default behavior of *Maple V Release 3*.

With **interface(errorbreak=2)**, input stops after any error is encountered, regardless of the type of error.

System and Language

Language

The following major changes have been made to *Maple*'s programming language.

Name Protection

R3

In *Release 3*, a facility for name protection has been added, which automatically returns an error message if you attempt to assign to *Maple* system variables.

```
> set := {1, 2, 3};
Error, attempting to assign to `set` which is protected

> lhs := op(1, x = y^2);
Error, attempting to assign to `lhs` which is protected
```

The protection facilities are available through the functions **protect** and **unprotect**. See the help page for these commands for more information and examples.

Global Variables

R3

A significant change has been made in *Release 3* to the way global and local variables are declared in *Maple* procedures. You can now include a **global** declaration statement, much like you previously declared **local** variables.

In *Release 3*, any variable in a procedure that is not *explicitly* declared either **local** or **global** is automatically declared **local** if:

- the variable appears on the left side of an assignment (:=) statement

  or

- the variable is used as an index of a **for** or **seq** statement.

Otherwise, in all cases, the variable is declared **global**. Whenever a variable is declared **local** automatically by *Maple*, a warning message is issued. You can control the display of these warnings with the **warnlevel** option to the **interface** command.

In previous releases, any undeclared variable was automatically declared as **global**. It is *always* wise to declare *all* variables in your procedures.

String Processing

In *Maple V Release 3*, strings in *Maple* are no longer limited to a length of 499 characters. Strings of *any* length are allowed. Two new and powerful commands, **SearchText** and **searchtext**, are also provided to efficiently search for a substring and return the position of the character that begins it. **SearchText** is the *case-sensitive* version of **searchtext**.

```
> s := 'Most Maple commands are programmed in maple.';
         s := Most Maple commands are programmed in maple.

> SearchText('maple', s);
                                39

> searchtext('maple', s);
                                6
```

Other Improvements

- The **anames** procedure now takes a parameter specifying which type of assigned variable is to be listed. Thus, **anames(t)** will return all names of type **t**.

- The function **combinat[combine]** has been renamed **combinat[choose]** to avoid conflict with the **combine** function in the standard library. The **trace** function in the standard library has been renamed **debug** to avoid conflict with **linalg[trace]** (i.e., the trace of a matrix).

- The command **makehelp** takes as its parameter the name of a text file, and transforms that file into a *Maple* **TEXT** object, which can then be used as an online help page within *Maple*.

- The command **numboccur** counts the number of occurrences of a subexpression in an expression.

- The **fortran** command now accepts an optional parameter **mode=str**, where **str** equals **single** (default), **double**, **complex**, or **generic**, to specify how function names are to be translated.

- The options **iris** and **wordsize** are now available to the **interface** command to query the current version of *Maple*, its user interface, and the number of bits used in your system's machine words, respectively.

System and Language

R3
- Use of **local** variable declarations are no longer allowed in functions defined with the **->** syntax.

R3
- The command **ssystem** has been added in *Release 3* for multitasking environments that understand the concept of a command shell (i.e., they are **not** implemented in the Macintosh and Windows/DOS versions of *Maple V*). **ssystem** allows you to execute system commands from within *Maple* and return the result to your session. Also, the maximum time spent performing the system command can be specified.

Index

-> syntax, 47
.m files, 41, 43
    converting, 43
? syntax, 4
?? syntax, 8
??? syntax, 8
@ and **@@**
    and **expand**, 20

algebraic fields, 18
    factoring over, 18, 19
algebraic functions, 18
    evaluation, 18
    integration, 11, 19
    solving, 19
algebraic numbers, 18
anames, 46
animate3d, 37
animate, 37
animation, 37
    three-dimensional, 38
    two-dimensional, 38
archives, 41
argument, 24
assume, 9
axes
    labeling, 35
    scaling, 33
    tickmarks, 35

Bessel functions
    integration, 11
Beta
    with complex elements, 24
blockmatrix, 25

ceil
    derivative of, 19
    simplification, 19
cfrac, 25
Chi, 20
color PostScript, 38
coloring, 36
    by function, 36
colspace, 22
combinat[choose], 46
combinat[combine], 46
combine, 17, 46
    logarithms, 19
    radicals, 16
companion, 25
complex variables, 22
    arithmetic, 23
conjugate, 24
continued fractions, 25
contour plots, 31
convert
    factorials, 19
csgn, 24
    derivative of, 19
curl, 27

data files, 42
data types
    complex, 23
debug, 46
DEC LN03, 38
derivatives, 19
DESol, 14
    operations, 15

INDEX

det, 22
diff, 15
differential equations, 13
    algorithms, 13
    exponential linear, 13
    numeric solutions, 20
    output forms, 14
    partial solutions, 14
    plotting solutions, 21
    series solution, 14
Dirac
    integration, 11
    simplification, 19
display, 35
display3d, 35
diverge, 27
documents, 5
    transferring, 5
dsolve
    numeric option, 20

eigenvals, 22
eigenvects, 22
elliptic integrals, 11
entermatrix, 25
erf
    with complex elements, 24
erfc
    with complex elements, 24
errors
    breaking on, 44
evala, 18
evalf, 15
    over complex floats, 23
example, 8
exp
    with complex elements, 23
expand, 12, 20
    and radicals, 16

factorEQ, 27
fieldplot, 32
fieldplot3d, 32
floor
    derivative of, 19
    simplification, 19
fortran, 46
fourier, 10
frac
    derivative of, 19
fsolve, 22

GAMMA
    with complex elements, 24
gausselim, 22
GaussInt package, 24, 27
gaussjord, 22
generating functions, 24
genfunc package, 24
Glchrem, 27
Glhermite, 27
Glnodiv, 28
Glsmith, 27
global variables, 45
grad, 27
gradient vector fields, 32
gradplot, 32
gradplot3d, 32
graphics, 5
    additions, 31
    animation, 37
    devices supported, 38
    improvements, 31
    output formats, 38
Greek symbols, 6
grobner package, 28
gsolve, 28

Heaviside

INDEX

integration, 11
simplification, 19
help, 7
keyword search, 8
topic browser, 7
hermite, 27
Hermite reduction
for integrals, 12
HP Laserjet, 39
HP plotters, 39

I/O
formatted, 42
ihermite, 27
lm, 24
simplification, 19
Imagen 300, 38
implicit plots, 32
implicitplot, 32
info, 8
input, 5
Int, 12
int, 15
integration, 11
across discontinuities, 12
interface
additions, 5
improvements, 5
interface
errorbreak option, 44
iris option, 46
prettyprint option, 6
version option, 46
warnlevel option, 45
wordsize option, 46
inverse, 22
invfourier, 10
invlaplace, 10
is, 9

isgiven, 9
ismith, 27
isprime, 27

jacobi, 27

keyword search, 8

language
additions, 45
global variables, 45
improvements, 45
local variables, 45
name protection, 45
laplace, 10
laplacian, 27
LaTeX format
worksheets, 7
Levin's u transform, 22
Li, 22
libname, 41
libraries
archiver, 41
multiple, 41
liesymm package, 25
lighting, 35
Limit, 12
linalg package, 22, 25, 27
linalg[trace], 46
line style, 34
line thickness, 35
ln
with complex elements, 23
local variables, 45
loglogplot, 33
logplot, 33

m2src, 43
makehelp, 46
march, 41

INDEX

mathematics
    additions, 9
    complex variables, 22
    improvements, 9
    numerics, 20
    packages, 24
    symbolics, 9
max
    simplification, 19
min
    simplification, 19
minimize, 15
multivariate polynomials
    factoring, 18

name protection, 45
networks package, 24
nthcover, 25
nthdenom, 25
nthnumer, 25
numapprox package, 24
numboccur, 46
numerical approximation, 24
numerics, 20
numtheory package, 25, 27

odeplot, 21
online help, 7
    keyword search, 8
    topic browser, 7
output, 5
    typographic, 6
output devices, 38
output formats, 38

packages
    GaussInt, 24, 27
    genfunc, 24
    grobner, 28
    liesymm, 25

    linalg, 22, 25, 27
    networks, 24
    numapprox, 24
    numtheory, 25, 27
    padic, 24
    stats, 25
padic package, 24
parse, 42
partial differential equations, 25
pffge, 25
piecewise, 20
plot data structure, 37
plotting
    ambientlight option, 35
    contours options, 31
    font option, 35
    light option, 35
    linestyle option, 34
    symbol option, 34
    thickness option, 35
    coloring, 36
    lighting, 35
    rendering, 35
    statistical results, 26
polygons, 37
polyhedraplot, 37
PostScript format
    color, 38
    encapsulated, 38
    worksheets, 6
printf, 42
printing
    window contents, 5
procedures
    creation, 43
Product, 12
protect, 45
Psi
    with complex elements, 24

INDEX

radicals
    denesting, 17
    rationalization, 17
    simplification, 15
radnormal, 17
randvector, 25
rationalize, 17
rcf, 25
Re, 24
    simplification, 19
Re, 24
readdata, 42
readline, 42
Regis terminals, 38
related, 8
Release 3
    marker, 1
    overview, 2
rendering, 35
restart, 43
RootOf, 13, 18, 19
    in algebraic functions, 19
round
    derivative of, 19
    simplification, 19
rowspace, 22

safeprime, 27
saving a session, 5
SearchText, 46
searchtext, 46
series, 15
share library, 28
Shi, 20
signum
    derivative of, 19
    value at zero, 20
simplification, 19
    radicals, 15

simplify, 12, 16, 17
    and radicals, 16
smith, 27
solving equations
    numerically, 22
special functions
    integration, 11
sqrt, 16, 17
    symbolic option, 16
sscanf, 42
Ssi, 20
ssystem, 47
statistical data, 25
stats package, 25
strings
    length, 46
    processing, 46
subpackages, 25
Sum, 12
symbolics, 9
system
    additions, 41
    debugging, 43
    error handling, 43
    I/O, 41
    improvements, 41
    libraries, 41

technical support, 4
text, 5
TEXT object, 46
textplot, 35
textplot3d, 35
trace, 46
transforms
    Fourier, 10
    Laplace, 10
trigonometric formulas, 19
trigonometric functions

INDEX

with complex elements, 23
trunc
derivative of, 19
type, 23
type checking
automatic, 44

unevaluated functions, 12
UNIX pic, 38
UNIX plotters, 39
unprotect, 45
updates files, 4
usage, 8

VCR controls, 37
vector fields, 32
vt100, 39

W
with complex elements, 24
Weight, 25
worksheets, 5
examples, 8
transferring, 5
Wronskian, 25

Zeta
with complex elements, 24